I'M JUST TRYING TO KEEP MY WIG ON STRAIGHT!!!

How Cancer and Chemo Suck

Cover Artwork by Jessica Noele DeWitt, http://jessicanoele.com/
Edited by Rosanne Welch and Karimah Gottschlack
Back cover photo by Joaquin Santana, Instagram:Icarus Moon
Copyright © 2016 by Dahlia D. Welsh
Artwork copyright © 2016 by Dahlia D. Welsh
Published by Dahlia D. Welsh, Island Flower
Instagram: WigOnStraight
Facebook: WigOnStraight
Twitter: WigOnStraight
http://www.wigonstraight.com

Library of Congress Control Number: 2010900328
ISBN: 978-1-365-67179-1
First edition, 2017
Printed in the U.S.A.

I'm Just Trying to Keep My Wig on Straight is my personal journey from cancer diagnosis to remission to cancer diagnosis to remission. I was diagnosed and treated successfully for liver cancer and Non-Hodgkins Lymphoma aka NHL within the same ten-month period. Yes, I've had cancer not once but twice. I am now in remission.

I'm Just Trying to Keep My Wig on Straight: How Cancer and Chemo Suck is not intended to provide medical information or help with a diagnosis. It is intended to provide a heads up to patients, caregivers, medical staff, co-workers, family and friends regarding what it's like to have cancer. Twice.

I'm Just Trying to Keep My Wig on Straight is my story, which was sometimes funny, sometimes sad, but always about living in the moment and living life to the fullest.

***

FORWARD

'Just put one foot in front of the other'. In a simple sentence Dahlia D. Welsh captures the essence of how to approach and beat cancer, one step at a time. *I'm Just Trying to Keep My Wig on Straight* is a comical, heartbreaking and heartwarming, practical journey through the eyes of one person from diagnosis through treatment to cure of not one, but two different types of cancer. There is no sugar- coating it, a diagnosis of cancer is life changing, anxiety provoking, and downright scary. As a tsunami of dreadful thoughts purge common sense from your mind, taking a step back to appreciate the spectrum of issues becomes essential. Dahlia has outlined a secret recipe for understanding each and every step in the process, and how each and every patient should think about that step, and make the best decision for them. For me, every patient's journey is unique; every patient brings their own individual characteristics, quirks, and behavioral issues to the journey. Some approach the situation with boundless anxiety, others with infinite skepticism, and some approach the experience with a very tongue in cheek, sarcastic sense of humor. Dahlia clearly embraces the value of humor. Though, while Dahlia and other patients may use humor to help them cope with their day-to-day struggles, the point in fact is that every patient can experience the full scope of emotions imaginable on any given day. These emotions can range the full gamut, from anxiety to frank fear, skepticism to doubt, many times ending in bouts of tears in the middle of the night. The unfamiliar of course will do that. What Dahlia has done for all patients in her book, is to tell them that these emotions are natural and normal. She transforms the unfamiliar, and provides patients and their families with the understanding necessary to demystify the unknown, easing their transition into the process. And she does it with deep insight, with a healthy dose of humor.

When it comes to cancer, everyone has their personal experience, their own ups and downs. However, for those who have emerged at the other end successfully treated, and possibly cured, every patient has told me 'it changed my life for the better'. Patients repeatedly tell me, 'I have a new appreciation for life', 'I no longer stress the little stuff', and 'I spend more time with my family, and I don't hesitate telling them, I love you'. You see, while a diagnosis of cancer is certainly life changing, it's not all in the wrong direction. Dahlia in her brave and inspiring book, shows us that emotions of anxiety and fear are normal and natural, and need not be paralyzing. She equips us with the tools that familiarize us with the unknown, and the personal experiences of someone who has entered the journey and come out the other end a champion of informed patient care. While that wig of hers now sits by her front door, it undoubtedly reminds her and us all, that there is nothing in life you can't accomplish with a bit of humor, a smattering of love, and a massive dose of positive thinking.

Owen A. O'Connor, M.D., Ph.D.Professor of Medicine and Experimental Therapeutics Founding Director, Center for Lymphoid Malignancies Columbia University Medical CenterNew York, N.Y.

Intro

Cancer comes in many forms and so does the treatment for it. Dahlia's book detailing her own experience with two bouts of cancer and her journey through the process of getting well is starkly honest and offers personal insight to patients and caregivers and to those who wish to be a support for those battling the disease. Her anecdotes paint a vivid and realistic picture of the trials one may face going through cancer treatment. Her matter-of-fact style dappled with humor shine as she openly shares her story. The raw details in this book are a reflection of vulnerability one inevitably faces from the time of diagnosis and throughout the process of cancer treatment. What's especially noteworthy is the courage and mental fortitude that was undoubtedly a big part of what buoyed Dahlia through treatment. While she admittedly does not pray, what was apparent to me even from the beginning of her second bout as she lay on my examining table in the moments before her breast biopsy was an undeniable spirituality, one that manifested during my encounter with her and to which she alluded to in the book. While for Dahlia it was very much a matter of putting one foot in front of the other and being able to laugh at times when it seemed like she couldn't even keep her wig on straight, it is so often one's willingness and readiness to connect with others and to acknowledge and share intimate truths that are a profound part of the healing process and almost universal among many patients going through cancer treatment. Dahlia's book is a wonderful example.

Steven Sferlazza, MD
Radiologist

Big Ups!

There are too many names on my medical team to thank everyone individually so, I'd like to give a mass shout out to Dr. O'Connor's office aka the best oncology team in New York City. The infusion center at Presbyterian Hospital and the myriad nurses who I met, who helped me along the way. The ER and ICU staffs at Presbyterian Hospital and Beth Israel respectively.

Thank you to my health team, yaaay!

Owen O'Connor, MD, PHD
Ellen Neylon, MSN, FNP-BC, RN, OCN

Robert S. Brown Jr., MD, MPH
Nicole Golden, NP

Vladamir Sheynzon, MD

Steven Sferlazza, MD

The Biggest Shout Out

Without my mom, Ilene Welsh, I don't know what I would have done. My mom is retired and should be chillaxin' and then I dropped this bombshell on her lap and she sprung into action. Now of course you're probably thinking well, she's your mom she's supposed to take care of you, which is true however, she knew nothing about cancer, cancer treatment and had to learn a whole new skill set just to help me survive day to day. My mom drove me to my appointments and treatments, which may seem insignificant but driving from Brooklyn to Washington Heights was a big deal for her and honestly it broke my heart to see patients going through cancer treatments alone. My mom bathed me, massaged me, fed me and climbed into bed with me when I was too tired to fall asleep yes, too tired to fall asleep. My mom was and still is my biggest cheerleader. She was the Shirley Maclaine to my Debra Winger in "Terms of Endearment" except with the happy ending! It's like she found some hidden strength that she almost completely exhausted nursing me back to health. I thank my lucky stars that I had my mama by my side every step of the way because cancer and chemotherapy suck!

I want to say thank you to my family and friends because they were truly there for me even when I lost faith that I would survive not the disease, but the chemotherapy treatment. And, the following people served as beacons of strength because of their own battles with cancer:

Ruth Brooks, my aunt who told me to never give up and helped me keep my eyes on the prize because she too had gone through cancer and was in remission.

Jean Roberts, I never met Mrs. Roberts but she is the mother of one of my college besties and although she's no longer with us I know the story of her struggle and I used to wear her wig caps to protect my scalp when I wasn't wearing my wig. It was like my special barrier against cancer.

Shirley Chude-Sokei, my aunt who lost her battle with cancer over a decade ago and unwittingly gave me a blueprint for how to deal with your diagnosis, treatment and outcome with grace.

Oswald Welsh, my Uncle Sammy who after battling cancer once said to me, "You know Dahlia, I'm ready to go. Everyone knows I love them, I have been lucky in my life to be surrounded by a great family, I have no regrets so, I'm ready to go". It broke my heart when he told me that but he was ready and we had the best times together in his final days. Blossoma Welsh, my Aunt B. is currently going through her own battle and our talks are absolutely wonderful because for once I have the upper hand in that I can show her the ropes. Aunt B. is doing just fine.

And last but never least Crystal Morris, Crystal unwittingly inspired me to write this book years before I even got cancer. Crystal went through cancer over 10 years ago and she's never had an easy time of things but her laughter is infectious and she told me this great story when she was first diagnosed. Crystal had gone to a family function and she was bald because of the chemotherapy treatments. Her cousin is a hairdresser so her cousin's five-year-old son knows a thing or two about hair care so when Crystal arrived he looked at her bald head with concern and asked, "Was it a perm?" to which she replied, "No baby, I didn't get a bad perm. I have cancer". We've laughed so many times about the inquisitiveness of children and their need to fix and understand things. *I'm Just Trying To Keep My Wig On Straight* won't help fix cancer but hopefully it will help you understand it a little better.

Cancer: 1
Dahlia: 0

LIVER CANCER

I was diagnosed with an autoimmune disease during my last year of college. Each year for about the last ten to 15 years I've had to get every scan known to mankind: bone density, CT aka CAT Scan, and MRIs with and without contrast. I've also had about four upper endoscopies to scope out esophageal varices. When your liver is cirrhotic you become prone to portal hypertension that can lead to bleeding. If during an upper endoscopy you are found to have enlarged varices your doctor will tie them off to lessen the chance of bleeding. I actually woke up during an upper endoscopy because I wasn't sedated enough. I choked on the tube down my throat and busted a blood vessel in my eye. Fun times. I also take Nadalol, which is a beta-blocker that keeps my heart rate down and decreases the chances of the varices becoming engorged with blood.

I've had about three or four liver biopsies. Liver biopsies allowed my doctors to actually see the state of my liver because they pull out a tiny cross section of it. Don't worry, the liver is the only organ in the body that can regenerate itself so what they took out has already grown back. FYI, you're awake during a liver biopsy because you have to inhale and hold your breath thereby moving your right rib cage up so that it doesn't block your liver. All these procedures were done over the years to monitor how cirrhotic my liver had become and how the cirrhosis was affecting my body.

I had been diagnosed with a chronic liver disease when I was twenty years old and a senior in college. When you've been scanned a bazillion times as I have been and you are always cancer free, you don't worry about it. Fast-forward 25 years later. I had filed the possibility of getting cancer in the back of my mind even though I knew that the course that my chronic autoimmune disease would take might eventually lead to cancer.

And so it did one fine summer day, June 10th 2014 -- the day before my birthday.

I had gone in for an MRI a couple months prior and didn't worry too much. I almost skipped the appointment because of problems with scheduling but then my liver team called because the scan showed the overwhelming possibility of me having cancer on three different areas of my liver. I have to admit that I was shocked because I did not feel like I had cancer. I mean I was tired at times and I was on a family trip a month before my diagnosis when my back became so hot I got the shivers. Could those have been some telltale signs?

Maybe? Perhaps? I don't know, however what I did know was that I had better go see the doctor my liver team recommended before my fears got the best of me.

My liver cancer, thankfully, started and was contained inside my liver. That means that it hadn't spread from elsewhere in my body to my liver or vice versa. As I mentioned, I was in the lobby at work when I got my liver cancer diagnosis. I hung up the phone and called my family. I considered leaving work but instead went back to my desk and back to work because cancer treatments weren't going to pay for themselves.

The next day my mom and I went to see the film *Maleficent* with Angelina Jolie. I usually try to go to a spa or take a trip for my birthday but I was glad that we already had something low key planned. Yaaay! You have liver cancer! Happy birthday to me!!!

When I met with Dr. Vladimir Sheynzon, my liver oncologist, he had a treatment plan in place and was absolutely wonderful and knowledgeable when answering my questions and addressing my concerns. My cancer was treated with chemoembolization. Radiation was an option but it probably would not have been as effective.

Chemoembolization was the preferred and most effective method to eradicate the cancer because the chemo would be injected directly into the cancer sites keeping the dosage low and local. Chemoembolization is an outpatient procedure and I didn't lose any hair since it was such a low dose of chemotherapy.

Was I scared? A little, I guess, but I try not to be a worrier I always try to keep things in perspective. More importantly I had who I thought was the best doctor for the job of cancer eradication!

On my list of do's and don'ts prior to the chemoembolization procedure was not consuming food or water and I had to shave my groin area. Of course being the procrastinator that I am, I waited until the last minute to do the last part. I knew the general vicinity my groin was in but wasn't 100% sure where to shave and for some reason it didn't dawn on me to just Google it. So in my hurry to get to the hospital on time I shaved everything eyebrows included -- just kidding.

During chemoembolization my doctor threaded a tube through my groin that would carry the chemotherapy drugs to the blood vessels feeding my three cancerous tumors. I was awake during the procedure. Trust me, I asked to be knocked out but that was not on the menu. Like with my prior liver biopsies I had to breathe in and hold my breath for what seemed like an eternity in order to keep my liver steady.

I had to go to the hospital and have the procedure done on two separate occasions. The first time they were able to treat two of the tumors. The last tumor was on the cusp of my liver. In order to not tax it too much and possibly cause liver failure they left the third tumor for another day. The first time I was treated I was given a local anesthetic. The second time, after the doctor was finished cleaning me up: removing the tube with the chemo drugs, cleaning the incision site where the tube was threaded, and administering pressure on the incision site so that it would not bleed. I told the nurse that I didn't remember when they had given me the anesthesia. She said they only give anesthesia if it's needed based on the patient's discomfort or pain level. What?! So, I went through a second round of chemoembolization with no anesthesia – not bad. To tell you the truth I wear my high tolerance for pain like a badge of honor.

When I went back the second time to treat the last tumor, boy, that was a long day. I hadn't eaten or drank anything all day and then, just as I was about to be wheeled into an operating room, here comes EMS with someone who had a stroke. That meant more waiting for me. My mom was there the whole time. Thankfully, she's very social and I think has a sort of senior form of ADD. She sat in the family waiting area, talking to everyone and relaying stories to me. She would go to the cafeteria for food, chat on her phone even though she wasn't supposed to – anything to keep busy.

The nurses that were on shift while I was there were very nice; they checked on me and kept me abreast of the situation. But heck, what can you do after you're in a hospital gown naked as a newborn and attached to an IV with liver cancer? I just waited it out.

Cancer: 2

Dahlia: 0

I'M SORRY I HAVE WHAT?

The path to my Non-Hodgkins Lymphoma diagnosis started with a lump on my right breast. A lump that was so visible it looked like I had a second nipple, it was pea sized and hard. I tried not to be alarmed because I have lipomas and I had just gone through liver cancer so surely I couldn't get cancer again so quickly, right? Right? Wrong.

I don't remember which came first: the discovery of the lump on my breast and subsequent mammogram or if I already had had a mammogram scheduled. However the appointment came about I'm just glad that I'm the kind of person who doesn't hesitate when it comes to health.

When I was at my appointment I knew there might be a problem when I was the last patient left in the waiting room of the mammography suite. I had already been examined and was waiting for the nurse to tell me I could leave. After what seemed to be an eternity the nurse took me into an office and explained that I needed to have a biopsy done of the lump.

I'm not gonna lie. My heart sank but I put on a brave face and went to work. I got the biopsy done the following week by Dr. Steven Sferlazza who was so wonderful. We chatted during the exam and he said that he had a Tom Jones song in his head but he couldn't remember the name until he saw my patient chart. It was the song "Delilah" and I'm Dahlia. I also reminded him that Tom Jones is Welsh and that my last name is Welsh. I was really hoping that was a good omen, that maybe the lump was really nothing to worry about, but no such luck.

The next week I was at work. You'd think I would've shut off my phone by now since every time I got a call it was bad news from a doctor's office. When my primary care physician called I was actually working as a hostess to a high level meeting. My doctor had received the results from my biopsy. It showed that I had cancer yet again, twice in the space of six months.

To tell you the truth, she sounded more disappointed about my results than I did or maybe in med school doctors learn the "I have bad news for you" voice. I don't know who I was trying to comfort but the basic gist of my response was, "Don't worry, I get bad news all the time. Thanks for calling, but I gotta get back to work" and with that I looked around to make sure my conversation wasn't overheard, slapped a smile on my face and kept working although what I really needed was a good cry and a big hug.

RUN DON'T WALK TO YOUR DOCTOR

When I was diagnosed with NHL, because of my liver disease – chronic autoimmune hepatitis -- my liver team immediately sent me to Dr. Owen O'Connor.

From the first day Dr. O'Connor was amazing. He sat down with my mom and me and explained exactly what Non-Hodgkins Lymphoma is and how the cancer works in the body. Although NHL is a very aggressive cancer it has a high cure rate. Dr. O'Connor made no promises that he could cure me, but he did have a plan of action for fighting the disease.

As time went on I realized that he was always open to changing the plan if necessary and I felt as though I was part of the decision making process. With any illness it is important to trust your doctor 100% and to have the kind of openness where you feel comfortable discussing any and everything. I would and still do ask Dr. O'Connor about treatments and possible dietary options to keep me healthy and he never hesitates to give me his honest opinion. He has been a leader in his field for decades and has been practicing for even longer and that alone gave me the confidence to literally put my life in his hands. And quite honestly if I said to him that I wanted a second opinion I wholeheartedly believed that he would have encouraged me to do whatever was necessary to feel confident that I at least had a fighting chance at beating this disease. So armed with our action plan, a lovely sketch he made of how NHL works, and his wonderful office staff, I was ready for the battle ahead.

SECOND OPINIONS

Second Opinions, I didn't get a second opinion because I trusted my oncologist, Dr. O'Connor and I didn't want to delay my care. I Googled Dr. O'Connor and found out that he was one of the top docs in the field of Lymphoma. But just as important to me was the fact that we had a great rapport from day one.

WHO TO TELL ABOUT YOUR DIAGNOSIS

In the beginning I didn't tell everyone about my cancer diagnosis because quite frankly I couldn't stop crying. I didn't want to keep talking about it over and over and over again because it made me sad. Beware because with some well-meaning people you find yourself comforting others although their intention is to comfort you. When you're dealing with something as serious as cancer you have to pick and choose who to tell, what to tell and when to tell.

If there is a time in your life to be selfish this is it. I needed to focus on myself and get as much rest as possible because even if I felt like a million bucks I knew that keeping the circle of people who knew my diagnosis small was the best thing for me. I shut down most communication with others because I physically didn't have enough energy to answer the phone, text or have visits.

As far as telling your employer, that's a personal decision. Even though it shouldn't, money often factors into every decision, including these dire ones.

R-CHOP

To treat my Non-Hodgkins Lymphoma I had six R-Chop treatments. I went to the hospital's infusion center for treatment every three weeks for about five months. The Rutuxin part of the R-Chop treatment is pushed into your IV via a big ass needle filled with red stuff. You have to chew ice while they slowly administer it into your IV line because there may be the possibility of you getting lockjaw. Really people? As if cancer and chemotherapy are not bad enough, let's throw in some lockjaw while you're at it.

My first chemotherapy treatment was on Christmas Eve. I had a half-day at work because of the holiday so at around 2pm I left and headed across town where I checked myself into the hospital.

Since my first chemo treatment was scheduled over the holidays I hoped that I would have more than enough time to recuperate before going back to work. Going forward, my treatments were scheduled on Fridays so that I would have the weekend to recover. The Monday morning after treatment I would go back to the hospital to get a *Neulasta* shot in my arm, which helped reduce my risk of infection by boosting my white blood cell count. Normally you should get the shot 24 hours after treatment but I was always wiped out the next day and because of my shady insurance I couldn't get the shot in Brooklyn, which meant another half day ordeal. First thing the Monday after treatment I woke up early, rushed to the hospital to get the shot and then rushed to get to work on time.

There are many ways that cancer is treated. My course of treatment was chemotherapy, no radiation and no surgery, just chemotherapy. The type of chemo I received was called R-CHOP and it was administered intravenously.

PORTS

It wasn't until after I started treatment and had to constantly get blood drawn and get plasma and blood infusions that the idea of getting a port came up. The nurses and phlebotomists who had to deal with my horrible little small rolling veins suggested it. I can't even begin to tell you how many times I was stuck with a needle during the course of my treatment and during emergency hospital stays but let me give you a hint, I was stuck no less than six times during one MRI visit. SIX! It was too much. This is part of why I was also mentally exhausted. After a significant amount of weight loss, being bald and having beaten up veins I seriously looked like a little crack head. Get ready to look at yourself in the mirror and not recognize yourself. It is a mentally jarring experience.

With all that said I still wouldn't have agreed to a port because from what I heard they can be more trouble than they're worth. You have to keep them clean and they don't always work. Also, I just didn't like the idea of a foreign object in my body. Mid-way through my ordeal I developed a routine to make injection days a little easier. I stayed hydrated and insisted that hot packs be put on my arms to help bring my veins to the surface.

MORE REASONS TO HATE CANCER

I mistakenly thought that since I was able to tolerate the side effects of chemo pretty well, I'd be able to go on with life as usual with, as I liked to say, a side of cancer. Wrong! As the treatments went on I grew weaker physically, mentally, and emotionally. Then, just before my sixth and last chemo treatment, I contracted sepsis, which totally KO'ed me. I could no longer work and ended up in a local hospital for ten days.

My New Normal

While your immune system is compromised avoid the following:

SHORTLIST OF CHEMOTHERAPY NO-NO'S:
- Manicures and pedicures because unsterile instruments may be used.
- Dental procedures including cleanings not only because of unclean instruments but because your gums may bleed – I had clotting problems due to my low immune system.
- Ob/Gyn examinations, which can be invasive due to scraping for tissue samples.
- I had a small lump in my left armpit that turned to be an ingrown hair which I had to let pop on its own rather than popping it myself.
- Shaving with a double-edged razor is a big no-no because you may cut yourself so use an electric shaver instead.
- I didn't use deodorant because I didn't want to take the chance of getting more ingrown hairs. If you have a cut under your arm from shaving and then use deodorant it may clog the pores. I had gotten to the point of being afraid of anything and everything causing an infection.

- Waxing any part of your body could bring infection due to unclean wax.

- Raw foods such as fish, veggies and fruit -- anything that is uncooked and may not be properly washed. Lunchmeat is included because the slicer may not be properly cleaned.

- Sexual contact came with a warning due to the possibility of pregnancy and the transmission of disease so it was advised to use two condoms. I laughed at the thought but my nurse was dead serious.

- Hot tubs/saunas because they are usually communal and if they are not properly cleaned you could be breathing in bacteria.

- Direct exposure to the sun-- if I was going to be out in the sun I had to wear sunscreen. Also, since I am dark skinned, overexposure after chemo would make me extremely dark. Now I love a tan as much as the next person but I didn't want to look supernatural.

PUBLIC TRANSPORTATION/BIG CROWDS

I live in New York City where the subway and buses are the best way to get from point A to point B. However over five million people ride the subway every day and they're not the cleanest places to be when you have cancer. They're jam-packed with people, many of whom are sick and don't take the spreading of their germs seriously. I could virtually see the germs coming at me in slow motion as people coughed without covering their mouths. I was lucky enough to have my mom drive me to most of my doctor appointments. She really stepped up to the plate and put aside her fear of driving to new and unknown places; that seemingly small gesture meant a lot. Honestly, I wish America was more like a lot of Asian countries where people who are sick wear facemasks so that they don't get others sick. If you were to wear masks here people would think you have the plague rather than the common cold.

USE HAND SANITIZERS A LOT

I always wash my hands when I get home or arriving at work but I wasn't big on hand sanitizers; however, during chemotherapy and remission, hand sanitizers are a must. I probably should've used my hand sanitizer more when going in and out of doors and when I had to ride the New York City subway, which really is a cesspool of nastiness especially for someone with a compromised immune system. Also, it didn't help that people were sneezing, coughing and touching everything in sight. In hindsight, especially since I was ill during the winter, I should've ceased shaking hands with people and blamed it on having a cold; that way I wouldn't have looked like a crazy germaphobe.

Surviving Overnight Hospital Stays

As anyone will tell you hospitals are some of the germiest, diseased places on Earth but you gotta go for treatment. During the course of my treatment I think my hospital stays totaled about two and a half weeks. I really wish I was more prepared and got the following advice:

SICK BAG

I kept telling myself that I would be proactive and put a bag together filled with necessities for possible overnight hospital visits. However, being a world-class procrastinator and believing that if I didn't pack an emergency bag I wouldn't need it, I didn't get around to doing it until the end of my treatments. You'll need a sick bag for different occasions like for overnights, hospital visits for infusion days which normally take anywhere between five and eight hours, and going to doctor appointments – the wait times can be brutal. It was always good to have bottled water, snacks, reading material, cell phone, computer, headphones and chargers. Include Afrin to help stop nosebleeds, tissues for aforementioned nosebleeds, balm for chapped and sometimes bloody, peeling lips, a sweater or blanket because hospitals are as cold as meat lockers, and an inflatable neck pillow. I was also given a digital thermometer at the beginning of my treatment.

HOSPITAL ROOMMATES

Get ready because unless you have the big bucks to stay in a private suite you'll probably have a roommate. I was lucky at times and had a room to myself when the hospital wasn't full and of course when I was in the Intensive Care Unit or I.C.U., but I had one roommate who talked/yelled at the top of her lungs the whole night. I pulled a Shirley Maclaine in *Terms of Endearment* move and insisted I needed to be moved because I couldn't sleep. I was moved the next day.

I had one roommate that turned out to have the flu – really? Um, did I mention that my immune system is shot to hell? It was Mother's Day and I got a nosebleed that just wouldn't stop so I had to go back to the hospital again. When I got home from my overnight stay I got a call from my attending doc and my heart sank. I was sure I needed to go back to the hospital but luckily enough they prescribed medication that either worked or I didn't get the flu. FYI, prior to the start of my ordeal and treatments I had gotten my annual Flu shot but apparently chemo knocks it out so it's no longer potent.

I had one roommate that didn't have cancer but she did have a bone disease, which they treated with chemo to keep at bay. Unfortunately, she didn't have siblings so stem cells were not an option for her. It's good to know that my sisters are good for something – hugs, kisses, bone marrow. I also had a roommate who was a lovely woman with a large, loud family. It was amazing how many people they could cram on her side of the curtain.

HOSPITAL BEDS

Special shout out to hospital bed companies because apparently they were designed to ensure you get the worst night's sleep ever. First of all nothing is as comfortable as your own bed. Secondly the hospital beds I slept in tried to 'readjust' every time I moved to make me comfortable, which was annoying not only because it was noisy, but because it was wholly unnecessary to keep shifting. It drove me crazy because the bed never got it right. I was never comfortable. So, kiss getting a good restful sleep goodbye.

IV MACHINES GOING OFF

Oh my God, my IV machines always seem to start beeping just as I got comfortable and was about to fall asleep. Beware that if you somehow tangle up your cord or bend it so that your IV fluids aren't able to flow freely, your drip will not only be slower and prolong your stay because you have to get all the prescribed meds, blood, plasma etc., but it will set off your machine. A low battery will also set off your machine so make sure that it's fully charged for those times when you have to unplug it and take it with you to the bathroom. Also, have the IV tree on the side of your bed where you can get most comfortable, meaning it shouldn't be on your left side if the tubes are in your right arm because you have to lay on your back during treatment and hospital stays.

MOM DID ALWAYS SAY TO WEAR CLEAN UNDIES

When I was first diagnosed with Non-Hodgkins Lymphoma, I convinced myself that I was going to bring sexy back to cancer. I was going to stay cute and polished and pulled together. However, most if not all of that went out the door when the tiredness, bloating, hair loss, loss of appetite, black fingernails and mental strain from cancer took center stage. After a while I couldn't care less if my skin was moisturized or if my wig was on straight!

However do heed your mother's words because as I got sicker and had to go to the ER more and more - everything seemed to start with the ER - I realized that there are some cute doctors working the late shift. Everyone needs some eye candy and during one visit I had actually heeded my mom's words and made sure I was lotioned up and wearing my cute gingham undies with the frills around the edge. Thank goodness I did because while I was in the ER as they checked my temperature, drew blood and ran IV lines, they also wanted to do a quick test to see if there was blood in my stool. Dr. McCutie told me what he needed to do and then gave me some privacy. I rolled over onto my side, pulled down my pants and undies just enough, Dr. McCutie came back behind the curtain, put on a glove, put his finger up my butt and it was that quick. Dr. McCutie took off the gloves, told me he would get the results and with that he was off to his next patient. All I could think was, *call me!*

INFUSION DAY CHEATS

Tip #1: Arrive at the infusion center to beat the crowd and get registered in the system as soon as possible. Anything that has to be ordered day of will prolong your appointment; for example, a test may have been left off of your requisition or you may need blood and/or plasma – all these "extras" need to be first approved by your doctor and then transmitted to the hospital to be fulfilled.

Tip #2: Depending on which facility you go to for treatments you can request a bed. Otherwise you might get your treatment sitting up in something akin to a lounger that can be comfortable but not as comfortable as a bed especially since some treatment days could last up to ten hours.

Tip #3: Make sure that you get a typecast no more than 24 hours before your treatment otherwise waiting for the results may prevent you from receiving treatment. What is a typecast? Well, in layman's terms, before every infusion of blood products they have to check your blood type before ordering blood or plasma. Now you're probably thinking the same thing I thought, why do they have to check my blood type every time? My blood type is not going to change. It won't, but due to hospital protocol they still have to check each and every time.

Tip #4: If you're particular about what you like to eat bring your own food and snacks. I had no dietary restrictions. However, if soggy tuna sandwiches or lunchmeat is your thing you should be all set. To tell you the truth, by the time I got my first dose of Benadryl I was out like a light for the duration of the visit.

Tip #5: Get your parking ticket stamped. The stamp didn't cover the charge entirely but it did help offset the cost.

Tip #6: Hopefully your caregiver is as gregarious as my mom. She really kept herself busy socializing with others in the waiting room, and with the nurses. She'd even take walks around the neighborhood of the hospital when she wasn't doing her crossword puzzles, which was fine by me.

And with cancer as with life, always keep well hydrated by drinking a lot of water. Ya gotta keep those veins plump for all those injections!

Sh*$ Happens And
Other Side Effects Of Chemo

DIARRHEA CHA CHA CHA!

It's pretty bad when you're sick from the copious amounts of chemotherapy coursing through your body and then on top of it your stomach takes on a life of its own. Whether you have diarrhea or constipation, what used to be the norm is out the door and has to be taken very seriously. Tell your doctor about your excessive "movements" because you never know if changes in your body's internal ecosystem could be a sign of bigger issues. For example, excessive urination can be a sign of a kidney problem or you may have an infection or a bug that is making you sick.

Follow your doctor's orders to the letter. You have to let your oncologist know what's going on because you have to help them help you. In one particular case I was constipated and worried for days that I might be taking a turn for the worse. Then I worried I had waited too long to say something. When I finally told my nurse about my bowel issue, the solution was as simple as taking Metamucil to get me going again.

MY LIPS

After every treatment the skin on my lips became very thick and by very thick I mean they came off in big chunks almost like fingernails.

BLACK JELL-O

Black Jell-O is the best way I can describe my bowel movements when I was in the hospital for sepsis. I was in the I.C.U. I was so bloated from ascites, a condition related to my liver disease, and all the I.V. liquids I was infused with. I could barely eat and I didn't have a bathroom so I had to use a bedpan 24/7. All the meds and liquids that were used to save my life made my bowel movements look like black Jell-O.

I would like to give another special shout out to nurses because they not only wipe away tears and running noses but everything else that runs. And they do it without complaining and immense helpings of understanding and empathy. Love you all!

PERIPHERAL NEUROPATHY

After six R-Chop treatments and two CT scans that came back clear of cancer, I was ready to rest and heal so that I could get back to my normal life. But another side effect I experience to this day is peripheral neuropathy, which feels like having frost bite on my fingertips, toes and pads of my feet. It's like those extremities have gone numb. You need your toes to walk and when it was bad I had trouble getting up if I was on the ground or a low position in yoga. The numbness has lessened but still lingers, and for some people it can get so bad they have to walk with a cane or walker to help keep them balanced.

CHEMO BRAIN

Chemo brain is a lot like pregnancy brain I've been told. I noticed my increasing inability to remember things and at times I had trouble focusing enough to get out coherent sentences. There was a delayed response when I answered questions and although I knew what I wanted to say the words took a while to come and were often jumbled. It's definitely one of those things you just have to laugh about. On average it's supposed to take about three months after the last treatment to clear the fog.

EYES

After my first chemotherapy treatment I saw cancer bubbles. It was like I could see the actual air particles in front of me. Also, after each R-Chop treatment my eyesight would go wonky for a couple of days. I had trouble getting them to focus, which was due to the high (100mg) dose of the steroid Prednisone you have to take for five days after each treatment. My eyesight is near perfect so I wore low-grade glasses to help me focus and that did the trick.

HOT AIR

I mean more than usual, pah rum pum! During treatment I wasn't gassy or burping a lot. Air would just escape from my mouth. It was the weirdest thing but not odd enough or consistent enough for me to be alarmed to the point where I told my docs. In hindsight maybe I was deflating physically and mentally.

THAT FUNNY FEELING IN MY HEAD

I'm not sure if there is a technical name for this but right before I would lose locs of hair I would feel surges of energy. My scalp would tingle and feel tight and hair would fall out and when I was completely bald the sensation stopped.

THE BOTTOMS OF MY FEET

The more treatments I had the darker the soles of my feet became. It was almost like I was being filled from the bottom up with chemo. However, since I'm in remission they have gone back to their normal color.

DRY MOUTH/MOUTH SORES/ORAL THRUSH

Dry mouth and mouth sores weren't really a problem for me but in case you get it, there is something called Magic Mouthwash that your doctor can prescribe.

TASTE BUDS

My taste buds dulled after every treatment of chemotherapy so I knew not to buy a lot of groceries or to make plans to eat out because I probably wouldn't eat the food at all or much of it. However, since I had three weeks in between treatments my sense of taste came back around week two after my treatment.

NAUSEA

I never experienced nausea, which surprised me because I get really bad seasickness. It's too bad that I didn't know earlier how great Dr. O'Connor was when it came to chemo dosage because I never used any of the very costly drugs I bought to combat nausea.

KEEPING TRACK OF YOUR MEDS

I've been taking medication for over twenty years due to my liver disease so I have probably used every kind of pill reminder case known to mankind. However, going through cancer was the most pills I've had to take and the most varied list that had to be taken at different times and different doses. I bought a large seven-day case that was broken down into morning, noon and night and I used my pharmacy's app for refills. The app also helped because during the course of my treatment every doctor I encountered needed to know what meds I was taking, the strength and the dosage and the app allowed me to give accurate information.

NOSEBLEEDS

You can get a nosebleed for a few reasons. I had a low platelet count so my blood had trouble clotting. I usually get nosebleeds once a year during winter, maybe because the air is drier, so my solution was to use a humidifier. You also have to be careful not to pick or blow your nose too hard. For immediate relief I would do the old standby of pinching my nose, stuffing tissue up my nose and using Afrin spray – not prescribed by my doctor.

HIGH POTASSIUM/LOW POTASSIUM

So, there I am in the Gap clothing store making a return when I get a call from a doctor – you'll get many calls from doctors. Good luck remembering them all. This was the after hour doctor calling because the blood test I had taken in the morning showed that I had a high level of potassium in my body. Who knew having too much potassium in your system could be potentially life threatening? FYI, high amounts of potassium can slow down your heart rate and weaken it. I finished my transaction and headed uptown to the ER crying the whole way. All I could think was I had just had my first treatment and already my health was beginning to deteriorate.

I was on the subway platform, surrounded by a sea of people. I kept my head down so that no one could see me cry. I called my mom and when I couldn't reach her, I called my sisters who live down south so they could reach her and let her know I was heading to the hospital. I felt awful. It would be one of many WTF moments. And you know, you can do many things in New York City but crying on the subway apparently is the one that will get you ignored the quickest.

RESTING

Even when you feel up to it, rest. Even when your caregiver doesn't listen and has people over to visit and wish you well, close your door and rest. By rest I mean do nothing just lay in bed and sleep. No Facebook. No texting or going on the internet. Don't make or accept any phone calls except from your doctor. Rarely do you have a really great reason to lay in bed all day so do it. That way your body has time to recuperate and either get ready for the next round of chemotherapy, heal while you're in remission, or strengthen so you can go to work. No one ever won an award for working when they were dealing with cancer – a lesson I learned the hard way because I probably should have put my health first a lot more.

HAIR LOSS

I've worn my hair natural meaning not permed or straightened for almost twenty years so the world of wigs, extensions and weaves was something very foreign to me. But I knew I had to go wig shopping because sporting the bald look at work was not an option. I went to a place that a friend recommended near Madison Square Garden. It seemed to take forever because I really just wanted my own hair or something that looked as close to it as possible. It was after my first chemo treatment and my hair hadn't started to fall out yet. Unbeknownst to me the store had a limit to how many wigs you could try on but the woman helping me let me try on a few extra. I guess she could tell that I was going through something tough. At times I felt like you could almost "see" the weight of the disease. I settled on a curly wig and then, at home I tried it on and cut it into a short curly style. Now if I could only keep it on straight!!

Within a week I went wig shopping again. I hadn't started wearing the first wig yet because my hair hadn't started falling out. However, the time came when I knew I had to cover my hair because my locs began to come out in clumps. I actually had a ritual every morning before work. I would bend over and shake my hair as much as possible so any of the loose locs would fall out. Then I would pray that no locs would fall out during the workday. It worked for a while until I began to lose hair in the front and there was no two ways about it, I had to wear a wig because no amount of headbands or bobby pins could stop the inevitable. My mom and a family friend took me to a local wig store. I thought it would be nice to have more than one wig and my sister Lisa paid for it which was a nice surprise.

As I sat in the chair at the wig store trying on different hairstyles my hair started to fall out like rain. And I lost it, I looked at myself in the mirror and wondered if I could get through treatment. Cancer is a disease, not a punishment, but it was another WTF moment. My mom tried her best to comfort me but there isn't much anyone can do to make you feel better. I will say that the woman helping me fit and style my new wig said something that may seem harsh, but actually was the right kick in the butt I needed. She said, "You're not the first person to go through cancer and you won't be the last." A little harsh, yes, but that's how we Jamaicans are – straight up with no chaser.

I was prepared to lose my hair but to me that wasn't the scary part. The scary part was how I would lose my hair. My mom offered to take me to a barber to have my head shaved but I didn't want a big production especially in front of a bunch of strangers. That day when we got home from the beauty supply place, I went into the bedroom and my sister Sophia called to see how I was doing. I cried so hard trying to explain that my hair was coming out in clumps and that I was getting bald spots that couldn't be covered. She told me to just cut off my hair, that I needed to stop torturing myself waiting for each lock of hair to fall out. I respond best to tough love.

When we were done talking I stood in front of the mirror and actually pulled the hair out of my head. My locks came out of my scalp without any resistance and then soon enough I was standing there bald as a new born. My mom came into my room and gave me a hug. All the stress of waiting to go bald had been lifted.

DON'T TOUCH MY HAIR

I enjoyed my coworkers. They were a great group of people and they unwittingly helped keep my spirits lifted. However, those first few weeks I showed up at work with my new wig were so stressful because of the amount of questions some co-workers asked – I tell you it was borderline harassment. "What did you do to your hair?" "Why did you change it? Is it a perm?" "Did you take out your dreadlocks and then curl it?" "So is it a perm?" Learn how to take a hint if someone says, "Oh, I'm just trying something new" or if they are obviously trying to end the conversation by walking away mid-sentence, then they probably don't feel comfortable discussing the topic. I had to tell one co-worker to stop asking me about my hair and stop trying to touch it. I am not a science project to be poked and prodded. I was bald as a baby's ass and I didn't have a way to firmly secure the wig onto my scalp so it only added to my stress level to have people asking about my wig and trying to touch it. All I could think was please leave me alone!

WEIGHT LOSS

Other than my hospital admission because of sepsis, I didn't lose any weight. And now that I am at my one-year remission mark I'm back to my normal size. I think I can attribute my lack of weight loss to my three-week cycle and low number of treatments. The first week after treatment my taste buds weren't really working. Then I had two weeks of normal eating before my next treatment. But don't think for one second I didn't try to score some medical marijuana. Unfortunately, if I was prescribed marijuana it would have been in pill form and not a big fatty that I could smoke, so what's the fun in that? Not that I would know, allegedly.

DIURETICS

My mom took me to a local hospital when I was treated for sepsis. I never anticipated blacking out for nearly twelve hours, which is what happened, and I hadn't given her the information she needed to get me to my doctors. At my local hospital, since they didn't know my complete medical history, they flushed my system with so much fluid I swelled up like a balloon. My circulation was so bad that they put circulation booties on my feet that would swell up to apply pressure on my feet and legs. Good luck getting a decent night's sleep with that thing compressing and decompressing with air every couple minutes. After I was released, my doctors prescribed a special diuretic cocktail of Lasix and Spironolactone that had me running to the bathroom every five seconds to pee. It was very difficult because I had put on about 45 pounds of water weight. I could barely move on my own while in the hospital. Once at home, it was really hard to get out of bed and to even walk on my own. I tried my best to anticipate when I had to roll off my sofa or out of bed so that I could rush to the bathroom so that I didn't wet myself. If you find yourself in this position I would recommend either a) not leaving the house b) staying close to a bathroom and/or c) wearing an adult diaper which didn't really work for me. I guess the one good thing that came from the , is that I basically urinated my way back down to what I like to call my "fighting weight", my weight back in college, which was around 125 lbs. I thought I looked crack head chic, but family and friends were not impressed .

THE UGLY CRY

After my chemo treatments were done, I was at a check up where I had to be scanned for the possible recurrence of cancer and the MRI suite had a waiting room full of patients with appointments, but only three machines and one that was being used for research students. Are you kidding me??? At the time I was still in a constant state of being tired and what should have been a three-hour appointment became a whole day affair. I was fasting for the tests and for someone with small rolling veins not drinking water for hours on end makes finding a vein even harder.

I was finally brought back to a small room where my glucose level was tested. Having not eaten for a good eight hours prior made my sugar level plummet, which was not good. I was given apple juice and it took three pinpricks on my finger to determine that my levels were finally acceptable for treatment. Then the nurse who was starting the IV stuck me with a needle in my wrist TWICE so that they could administer the contrast during the procedure. She did such a horrible job the needle "thread" was literally hanging on by a wing and a prayer in my vein. So it had to be redone because if they ran the contrast it could leak into my body, which would be very painful.

But wait, it gets better. Since it was so hard to start a line they suggested that maybe I come back another day to have the MRI because the amount of time it was taking them to test me was backing up other appointments. Come another day? I don't think so. They would have to pry my cold dead body from the machine before I let that happen. Another nurse pricked me another two times trying to get the line started and so six pin pricks and three nurses later the line was properly run. By the time the test started I was a crying mess from the pain and frustration but I finally got it done. Note to self: Confirm whether you can consume water prior to scans. I think for the good majority of them you can.

REMISSION VS. BEING CANCER FREE

I thought that after your chemo treatments were done you were in remission and cancer free. Imagine my surprise to find out that I will not be cancer free until I am 10 years out with no incident. Every year in remission without an incident is a pretty good indicator that all is well but relapse will always be on my mind – or worse, a third cancer of a different variety.

When The Cancer Is Gone
And It's Time To Move On

DONE WITH CHEMO, NOW WHAT?

I remember my first appointment after I was done with chemotherapy. My doctor and I discussed next steps and that I wouldn't have to come into his office as often for appointments. Now trust me, from my first diagnosis I was looking forward to the day of being done with treatments. I looked forward to not having to go into the office for checkups every two weeks. But when the time between checkups expanded and they said, "See ya in two months", I went into panic mode. My doctor and his wonderful staff – NP Ellen Neylon and the office staff, Erica, Joanne, Kathleen, Michael and Emily – were such an integral part of my treatment and daily life it was like being at a family reunion and having the first flight out. Now don't get me wrong, I understand that not seeing them as often meant that I was on the road to recovery but it was an emotional ding that I hadn't expected. My elation didn't last long because true to form, my immune system was still weak and I developed a new condition called Thymic Rebound. My white blood cells weren't replicating like they should've so I had to have a bone marrow biopsy and make subsequent trips to the infusion center for more Neulasta shots.

REDEFINING MYSELF

You know what really sucks about cancer? For me it wasn't having cancer or even the diagnosis but the treatment. Chemotherapy tears you down, strips you bare and kills everything in its way and then leaves your body to rebuild. Hopefully your body is strong enough to not only withstand the initial attack but also the treatment.

When I was done with my treatments and it was time to get back to normalcy, I didn't know what to do with myself. I mean I was so used to moving slowly, having chemo brain, not having an appetite, monitoring my stool, checking my temperature, getting blood drawn, getting immune boosting shots, sleeping with my phone in my hand so that I never missed my doctor's call and going to my doctor's appointments that I did not know what to do when it was time to heal. When my NP Ellen told me that they wouldn't need to see me for another month I was sad because they were such a great and tremendous support system. Although I missed seeing them, I knew that unfortunately it was now someone else's time to be treated. Cancer sucks. It really does, no doubt about it, but it also brings really great people into your life like my healthcare practitioners.

And you can't deny that it changes you in many, many ways. Among them:

CHEMO HAIR

Yep, that's what I call it and I'm sticking to it! I used to have dreadlocks like Bob Marley. They were long and thick. Now I have soft curls without the thickness and tinsel strength of my natural hair. Sometimes I love it, sometimes I don't, but I like change so this new hair signifies a new phase in my life – the post cancer phase.

LAB WORK

I have to keep on top of my lab work because it's the best way for my medical team to monitor me now that I don't have weekly visits. Having a liver disease prior to cancer means that my immune system struggles to get to a normal level; for example, my platelet count hovered around 18k-20k while a healthy platelet count is 150k to 450k.

ALKALINE WATER & SOUR SOP JUICE

I read online that cancer cannot live in an alkaline environment so I bought an alkaline water pitcher because I don't have four grand to spend on a kitchen sink top unit. Also, Sour Sop is a fruit native to Jamaica and a lot of Caribbean countries. In Spanish it's known as Guanabana and it's supposed to also have cancer-fighting properties. I am not religious about anything so I don't consume the aforementioned on a regular basis but hey, a placebo effect to keep cancer away is better than no effect. I also upped my daily dose of green leafy vegetables especially kale, and for me that meant just eating them. Prior to treatment the only thing green in my diet were jellybeans. So kale it up!

SWIMMING/PUBLIC POOLS

After my treatments as my hair slowly grew back, I went to a water park with my nephew. I didn't think ahead and wore my wig because I was self-conscious about being bald. In New York City sisters rock naturals all the way down to their skullcaps but down south it's a very different story. Also I had an unfounded fear of things falling on my baldhead and cutting my scalp before getting into the water. I went to the bathroom, took off my wig, and put on a swim cap. I was barely two months into remission, but I had to live my life and enjoy spending time with my nephew so not even a baldhead was going to stop me!

THYMIC REBOUND HYPERPLASIA

Chemotherapy will wreak havoc on your immune system and to this day all of my important counts are low; Absolute Neutrophil Count (ANC), White Blood Cell (WBC), platelets. After I was in remission for a couple of months, one of my checkup scans showed that I was experiencing Thymic Rebound Hyperplasia – say that five times fast. I tell you the hits just never stop coming with this disease. The best way I can explain it is that up until puberty, your thymus gland grows and then it stops and shrinks. During chemotherapy your thymus gland thinks, 'hey, it's time to start growing again', hence the rebound affect. The thymus gland is a primary organ of the immune system. It normally shrinks back to normal on its own after chemo. I had to be monitored, which occurred during my regular checkups. It is not normally seen as a cancer relapse but as a non-malignant rebounding.

BONE MARROW BIOPSY

And just when I thought it was safe to go back into the proverbial water... Several months after my last treatment my white blood cells were still not up to par so I had to get a bone marrow biopsy to determine what was wrong. Now I am not adverse to pain. I've gotten two wisdom teeth pulled without anesthesia and I even woke up gagging on a tube down my throat during an upper endoscopy but having a bone marrow biopsy took the cake.

My procedure was performed by a research fellow who is also a doctor but more on the research side as opposed to the clinical side. He hadn't performed as many as a specialist so there was another supervising doctor in the room. It is an outpatient procedure so my mom was there as well. She kept asking the fellow if they were going to do a spinal tap and he assured her that they weren't. Prior to having a bone marrow biopsy I highly recommend emptying your bladder. To prep for the procedure, I pulled my pants and undies down to just about the top of my butt crack. The doctor tucked what sounded like a public toilet seat cover into the top of my pants, and there was a hole exposed where the procedure would take place. He numbed the area with a local anesthetic that I think is designed to not only numb your butt, but also the area near the bone – I don't think you can numb bones. He then used what I am assuming was a corkscrew device to extract an actual sliver of my bone. Imagine someone using a corkscrew to take the cork out of a wine bottle – this was the same thing with a lot of pressure. The doctor then used the needle to extract marrow from my bone and oh mama that's where the pain came in.

In their defense I was warned that the marrow extraction would hurt but I actually whimpered. The pain doesn't last long; however, let me tell you it is impressive to say the least. When the procedure was done, the doctor wiped the area with a cotton swab and put on a bandage. I rolled over onto my back to use pressure to stop the minimal bleeding I experienced. The anesthesia typically wears off in about an hour. I didn't experience any pain when it did. In the event of pain, it was suggested that I take half an Oxy from my stash. I balked at the idea of possibly becoming hooked but was reassured that I wouldn't get strung out on a half an Oxy.

EXERCISE

Don't let anyone, including yourself, push you into doing more physically than you can handle. Caregivers, family and friends mean well but they don't have cancer – you have cancer – and they have no idea what you're going through. In the morning I would feel pretty strong but sometimes within two hours I would be wiped out. People only want to see the energetic side because they want you to be well for you as much as for them. Going up and down stairs, stepping off curbs, and even just walking down the block takes a lot of energy, but people don't understand that and think that if you just do this or do that you'll recover faster. You have cancer, not a broken toe. No amount of exercise will allow it to work its way out of your system, so say thanks for the advice and I'll be here watching *The People's Court* when you get back from your walk.

THE CANCER CLUB

There is nothing great or special about being in the cancer club, nothing. You tell people you know who've been through it and they wish they could have denied you admission. You hear so much about cancer that it's mind blowing when you get your own diagnosis. I wish no one else had to become a member of the cancer club, but until there is a cure, the "membership" will continue to grow.

HOW TO KEEP YOUR WIG ON STRAIGHT DURING CANCER

Throughout treatment there were several instances where I almost lost my wig both literally and figuratively. I had cancer during the winter, so when I arrived at work I would take off my coat, put my pocketbook in my desk drawer, and head to the bathroom to wash my hands and take off my winter hat. Thank goodness no one was in the bathroom when I did because one time my wig came off with my hat!!!

I stood there looking at my 'lil bald self in the mirror. Let me tell you, if there was an Olympic event for jumping backwards I would have won the gold medal because I immediately leapt backwards into a bathroom stall and slammed the door shut. I fixed my wig cap, shook out my wig, and then hurriedly exited the stall so I could check to make sure it was on straight. Phew! Crisis averted!

There was also the time when I was on the train and felt my wig coming off my head! When you're already self-conscious about your "hair hat", any little thing can set off your imagination. There I was on the New York City subway with my wig falling off, so I tried to nonchalantly look at my reflection in the subway door. My wig was still on so why did it feel like it was slipping off? As it turns out it wasn't my wig making a run for the border, the wig cap I wore underneath my wigs was creeping up – phew!

I had just finished my last chemotherapy treatment and the weather was warming up so I had on my "summer" wig that has an auburn streak in the front. My mom and I were packing up since we practically moved in during my daylong treatments. The IV had been removed, bladder emptied, but the effects of the Benadryl they gave me for pain was still in my system.

Without a mirror, I shook out my wig to fluff it up and plopped it on my head. My mom gave me a strange look and then asked, "Isn't the part with the color supposed to be in the front?" Ha, why yes!

Another time I was at my desk at work and was stretching when my bracelet got caught in my wig!! Thank goodness I realized what had happened before I ended up pulling the wig off my head in front of my poor unsuspecting coworkers.

And lastly, after a day of literally being stuck with needles nine times. I went to change my clothes and my nurse came with me to the changing room to see if I needed any help. I have to admit I was feeling woozy after not having eaten in about twelve hours, including the pre-fast. I dressed and realized that there was no mirror in my room so I shook out my wig and put it on. I looked at the nurse and asked her with a bit of flair if my wig was on straight, and after busting out laughing, she assured me that I was good. I had to find the humor in things because that was who I was before the cancer and the person I wanted to be during and after treatments.

Know that most days you will feel fine because having cancer doesn't feel like anything. So it's the side effects of the treatment and your mindset that are the biggest obstacles.

FOR THOSE STILL FIGHTING

While I was going through cancer, I did feel a special kinship to others who were in the same boat. There were many high profile cases. The ones that really got to me were ones like Leah Stills, the daughter of Cincinnati Bengals player Devon Stills. Cancer is tough at any age but to be so young and dealing with a disease this big is mind-boggling. Leah's will and determination was inspiring.

Nothing is promised including today and tomorrow, but it's your approach to your todays that determine your tomorrows. When a whisper of hair finally started to grow back I remember brushing my scalp like I had a head full of hair and saying, "Look Ma, my hair is growing in." She came over and inspected my scalp, angling it toward the light and said, "Yes. Yes, I see it!" We had a great laugh. Everyone deals with life's greatest heartbreaks differently so take care of your mental and emotional self as well as the physical.

UNCLE SAMMY

My uncle Sammy succumbed to brain cancer many years ago, and, you know, the greatest lesson I learned from him during that time was that he was okay with his fate because he had lived life to its fullest. In the end he was surrounded by love. The second thing I learned was to make it unequivocally clear who and/or what you want around you. Uncle Sammy didn't want any crying or tears otherwise you would get the boot! And so I did the same however as it turns out discussing my condition and explaining my condition a bazillion times made me cry. I limited who I told and when I would tell them. I knew I couldn't tell everyone. I would have spent entirely too much time trying to assure them that I would be okay as they acted like it was the end of the world when only the end of the world is the end of the world.

I lived with my mom through this whole process and I was actually afraid to go back to my own place, but I wanted to give her some space and start to rely on myself more. Those first nights I was home I missed her so much, but as time went on and I got stronger, I was able to settle into my normal routine, which now included daily check ins and still staying with her from time to time.

Once you've gone through cancer twice you can't help but worry just a tiny bit that it'll come back for a third time. About six months after my last chemotherapy treatment and around the anniversary of the discovery of the lump that started it all, I felt yet another lump under my left armpit; this one disappeared on its own over time, phew!

MY AUNT SHIRLEY'S BATTLE

I was there through Aunt Shirley's chemotherapy treatments. I took her to doctor appointments, bought her food, reminisced about life and I also witnessed her symptoms from chemo. Symptoms that I would mimic nearly 15 years later like her loss of appetite, black fingernails and toenails, hair loss, and chemo brain.

Aunt Shirley always had such a good spirit and I often wondered if she was there holding my hand as I went through my own tough times with cancer. I don't know, but I tell you what, I remember Aunt Shirley whipping off her headscarves as soon as she entered the house. Man, nothing felt better than taking off my wig when I got home and rubbing my head and giving it time to breathe. It was such freedom; I could just be myself in my sanctuary. It's like coming home after a long day of work and taking off your bra – women know what that means. I miss my Aunt Shirley and if I must admit it I also miss her cat Bubbles, too.

Hmm, That's Good To Know

HOW TO HELP SOMEONE GOING THROUGH CANCER

Luckily enough I had people who wanted to help in any way they could. If you can't help with grocery shopping, making meals, going to doctor's appointments – or maybe you just don't know what to do – the best way you can help is by giving blood. Cancer patients need blood and plasma because the disease and the treatment weakens the immune system. We need your platelets and blood cells.

THE POWER OF SOCIALIZING

My friends did everything in their power to lift my spirit. One friend exclaimed via Facetime, "I love it" when he saw me bald, "Gata 2015!" Even though I was not feeling beautiful at all. (In Portuguese, to refer to a woman as *gata* means to call her beautiful.) My friends took me out to eat, to concerts, and once even hired a car service to take me all the way to Jersey so that I could have a mini-break. My friends are awesome!

Now I'm on the road to a year in remission. My side effects are lingering but subsiding. I am trying to stay focused and get my life back on track because I've always had hopes and dreams to fulfill.

TRAVELING WITH MEDICATION

Before taking a flight, make sure that you can wear your wig through security because some hair pieces have metal, like the hooks that secure it to your head, that will set the detectors off. Also, check with your airline to see if you can take your medication onboard because when you're traveling, liquid meds may put you over the top of the liquid requirements. I kept my liquid med Atovoquone in its original bottle with the pharmacy label, and security told me that was the proper way to travel with liquid medications and let me through.

GETTING AROUND TO DOCTOR APPOINTMENTS

In New York City there is a service named Access-A-Ride that helps sick people get around. I would suggest registering for every means of support so that you're prepared just in case your caregiver is not available.

MEDICAL DIRECTIVE

If you don't already have one in place, this is a good time to decide who should make life and death decisions if you become unable to make the decision yourself. I love life, I love traveling, I love socializing, I love being out with my friends and family, I love being by myself. I love just existing and if at any point in time I can't do that, pull the plug. Hospital beds are uncomfortable enough and lying in one indefinitely is not who I am. I asked my mother if she would be able to pull the plug and she said no, that maybe one of my sisters could do it. She wouldn't want to, but after much prayer, I think my older sister Sophia could do it and she agreed.

I DON'T WANT TO HEAR YOUR CANCER STORIES

Sorry, but while I was going through cancer, hearing about your cousin's uncle's best friend's wife who died from cancer was not exactly uplifting. The most frustrating thing about these conversations was that 99.99% of the time the person telling the story didn't know what type of cancer the person died from, never had cancer themselves, and didn't really understand that cancer comes in so many flavors. Factors such as age, when the cancer was detected, race, and socioeconomic background, often play in the type of cancer some people are susceptible to. So, sorry, but listening and keeping conversations uplifting go a long way. Better yet, silence is golden. With that said, the only caveat was in regards to close friends and other loved ones who lost people to cancer because it was therapeutic for us both to share those experiences together.

CAREGIVER BURNOUT

I am lucky because my mom is retired so she had the time to take care of me. However, caregivers get burned out, too, going to long doctor appointments, picking up medications, staying overnight in hospitals, going to ER visits, cooking meals, wiping away tears, providing moral support. My mom is my mom and will do anything for me, but she's human too. As best I could, I would try to be independent so that she could get a break, which is not something she was too happy about because I needed to rest, not put my health at risk because I wanted to protect her.

SAY 'YES' MORE AND 'NO' WHEN YOU NEED TO

I am so used to being independent and was raised to be independent so it was hard to accept help, gifts, favors, etc. However, dealing with cancer showed me the power of saying yes. The phrase "It's better to give than to receive" is true because if you don't receive by saying "Yes", others don't have the opportunity to help.

Now, here's how I got to yes. When I was up to it I started taking calls and texts, and letting important people know why I had disappeared from the radar. I didn't take the social media route with tweets or a shout out to my FB family.

Your loved ones really just want to see you with their own eyes, put their arms around you, and know that they're not going to lose you. Now, I readily talk about my experience with anybody because I feel that there is nothing to be ashamed of, which is how I initially felt. It's informative for myself and others, it's a great topic of conversation and, hell, I'm at the point where I can laugh about my experience.

I DON'T BELIEVE IN GHOSTS BUT THERE WAS THIS ONE TIME...

My maternal grandmother's nickname is Ms. Ella. My grandma was the greatest: she was sweet, round, and squishy, which was perfect for hugs and cuddling. She was short in stature but she was "tallawah" - Jamaican patois for strong and sturdy. My mom gave Ms. Ella a wooden music box with a ballerina on top that when wound up would play a song and the ballerina would twirl around. When Ms. Ella passed away, my mom kept the music box. Before I got my liver cancer diagnosis, the music box had separated from its pedestal and my mom asked me to glue it back together, which I did with super glue. Actually I used so much super glue it wouldn't play!

So, one night right after a chemo treatment for my liver disease, I was in my condo in agony from charley horses in my legs and stomach. I drank Gatorade and tried to massage my muscles, but it was impossible to even walk and I was reduced to tears when all of a sudden, Ms. Ella's music box began to play. I was able to calm down and relax as the pain eventually subsided and I drifted off to sleep feeling as if my grandmother was watching over me.

SPEAKING OF MASSAGES

I've found that massages, foot massages in particular, seem to help the lingering peripheral neuropathy in my feet. I also wear Ayurvedic slippers that have tiny wooden balls that not only massage but also target certain pressure points designed to heal certain ailments.

FINANCIAL HELP

Even if you think you don't need it, look into what financial aid may be available to you or have your caregiver do so. There are cancer organizations that may be able to provide help, but you have to be diligent in finding out which one is for you. For example, I went to a website and in the drop down menu neither liver nor NHL was listed. That's why it's also very important to donate to these different organizations even from a purely selfish standpoint. For example, if you donate to breast cancer research, liver cancer and NHL don't get some of the money.

Make sure that your doctor or doctors take your health insurance and don't let your pride get in the way of getting Medicaid, which is great for those who have gotten the wind knocked out of their sails. Personally, if I wasn't so worried about what people might think, I would have looked into help from the government earlier and saved myself about four months in premiums and coverage for all the medications I had to take. One medication in particular was initially quoted at $1,300 per month! I left it at the pharmacy because I simply could not afford it. However my doctor was able to get it for me – your doctor's office sometimes has to convince your health insurance the medical necessity of medications.

PHLEBOTOMISTS

When it comes to getting your blood drawn, find a nurse or phlebotomist who can find your vein the first time and once you find a good one, go back to them as much as possible. Also, pay attention to what gauge needle works best for you. I'm a small butterfly gauge and not a straight needle person; however, such a small gauge doesn't work when you get injected with contrast during MRIs.

I can only recommend two things that may help having IV lines started easier. The first is to drink tons of water, especially before an appointment to plump up your veins. Also, hospitals sometimes have a quick heating pack that they can put on your arms that will bring your veins to the surface. And if you do anything please heed my warning: DO NOT LET A DOCTOR DRAW YOUR BLOOD. It is not common practice for doctors to draw blood and run IV lines, so they don't have as much experience as nurses. They also seem to want to stick you in the oddest places. One doctor did a blood draw through a vein in my foot. Another one started an IV in my NECK. Now here's my thinking: if you're going to go start a line in a place that is so uncomfortable and scary for the patient, shouldn't you get it right THE FIRST TIME? Nope, this doctor didn't so I had to go through the experience on my right side and then my left side -- and then the piéce de résistance? The line wasn't even used. I had yet another doctor who wanted to use a vein in my groin, which I have since been told is EXTREMELY painful. So do your research on ports and know where you draw the line, because you do have a say in all matters dealing with your care.

WORKING WHILE ILL

I tried my best to continue working during my treatments. Although I felt drained a lot of the time, I still felt physically and emotionally able. Socializing with my co-workers helped to lift my spirits, and I had bills to pay. However, as my chemotherapy treatments progressed, rather than being able to anticipate the symptoms and tolerate them, my immune system got weaker and so did my body and spirit. I don't know how I was able to do it but I did push through, even in the dead of winter. I lived my "normal" life until I couldn't.

AVOID DRIVING

I found, especially in the immediate months after treatment, that I was too discombobulated to drive. I had intense chemo brain and peripheral neuropathy, so I just didn't feel I could focus properly behind the wheel of a two-ton vehicle, so I left the driving to others.

In New York City there are care services with transportation benefits for which you may qualify like Access-A-Ride and reduced fare Metrocards.

BEING OPTIMISTIC ABOUT YOUR OUTCOME

Dying was never an option that I considered. Could it happen? Yes. However, I just woke up every day and put one foot in front of the other. Cancer is hard on your body, your life, your mental faculties, and your emotions. It'll keep you up at night and take the wind out of your sails when you wake up in the morning and realize, 'Wow, I have cancer', but here's the 'but' – don't give up. You are stronger than you think. Take all the hugs, kisses, texts, tweets, prayers and love sent your way and use it as your energy source. Now is the time to put ego aside and say yes I need you, it, them, whatever it takes to get you through cancer and chemo.

PERSONAL HYGIENE

Once the lethargy that came with chemotherapy started, my mom asked me when was the last time I had a bath. Honestly, I was so tired I couldn't remember, and I didn't care to put in the enormous effort, so I told her I'd bathe the next day, which she wasn't having. She ran the bathtub, and as I slipped inside the warm water, it was the best feeling ever. I played some Aretha Franklin, Otis Redding, and Etta James, and had my spirits immediately lifted.

HEAD SCARVES

I got more emotional as the days passed. One of my college besties gave me scarves and a wig cap that her mom used to wear during her fight with breast cancer. The scarves gave strength and at my lowest points I would wear the wig cap as "protection". I would talk to Mrs. Roberts and thank her for her strength as I was going through cancer. It was very comforting; I don't pray, but I do say thank you.

WIG PRESCRIPTION

Depending on your insurance carrier, you may be eligible to be reimbursed for buying a wig while going through chemotherapy. I was with a health company that is now defunct. They made me jump through so many hoops for reimbursement that I literally didn't have the energy to keep trying to find a wig store. They provided a list of vendors that carried health supplies. That list easily had 100 companies, and there was no break down by those who carried wigs. So I had -- per their stipulation – to spend at least 15 - 30 minutes calling to find out who carried wigs. Most of the places only carried medical supplies, i.e. bedpans and walkers. And although the companies were in my zip code, Brooklyn is huge so how was I supposed to get to them if I couldn't even find a place that carried wigs? I asked the health insurance rep for a breakout of stores with wigs, and they couldn't provide it. Hopefully you will have better luck, and this is where having an advocate to do the calling helps. My mom was caring for me alone, and as my mother she was going through her own emotional turmoil, so I didn't want to burden her any further.

IN CASE OF EMERGENCY - I.C.E.

Put your doctors' and caregivers' information in your phone under ICE because if anything were to happen and you couldn't communicate for yourself, emergency workers know to look there for information.

I also would suggest making a copy of the business cards of all your doctors, and even the back and front of your medical cards; that way everything is in one place for easy access. I would hand out copies at every new appointment.

ONCOLOGY STAT PROTOCOL CARD

While undergoing treatment I was given an orange card that I carried in my wallet. I called it an ER card. The card alerted medical professionals that I was an oncology patient and from which hospital I was receiving treatment. I thought about getting a med alert band and looked into several smart jewelry type devices, but nothing on the market really enthused me. Also, because I was going through treatment for a relatively short period of time, I didn't think they were worth the investment.

HAVE A MANTRA

When times got tough I couldn't think straight. I couldn't move out of bed. I was bloated. I was overweight. I was underweight. My eyesight was bad. I was weepy. I was feeling hopeless. My fingernails and toenails had turned black. My eyes were red. The hospital beds were unbearably uncomfortable. My arms were black and scarred from the intravenous chemotherapy. My veins were so overused they developed scar tissue. The Benadryl knocked me out. The morphine made the glands in my neck swell. My potassium was up and then down. I could barely walk without help. I walked up and downstairs like a toddler – one step at a time. I couldn't eat due to lack of appetite. The bus rides into the city with the people coughing, being at work and not sure whom to tell what and how much about my condition, just in case. When my family came to visit from as far as Georgia and Jamaica and just before they arrived, I was admitted into the hospital. Through it all I had a mantra that my Aunt Ruthie told me. When she went through her bout with cancer years prior, she said, "I have cancer but cancer does not have me." Through the tears and the heartache and the longing for normalcy, that one phrase – along with the love, support, prayers, prayer lists, and calls – got me through it.

WORKING AFTER CANCER

Everyone should go back to work at their own pace; however, know that going through cancer is considered a disability. Finding out that I now had a disability threw me for a loop because I only found that out while filling out job applications. I don't know if that's a good or bad thing but now I had a whole new classification.

The incident that forced me to stop working started with a very high fever, which may or may not have been exacerbated by my use of an electric blanket.

HEALTHY IMMUNE SYSTEM

It is my non-medical, non-endorsed belief that my Non-Hodgkins Lymphoma diagnosis came because my immune system was compromised from having a lifelong liver disease.

While I was being treated for liver cancer the "back door" to my health was left open, thereby giving Non-Hodgkins Lymphoma a pass to come in and wreck my health.

CANCER PLAYLIST

You know how when you go to the gym or for a walk or on your way to work, and you have a playlist that either motivates you or calms you down? I had a cancer playlist that was about five songs long that I would play whenever I needed a little extra motivation to get up. This was my playlist:

- "Yesterday" by Mary, Mary
- "Amazing Grace" by Mahalia Jackson
- "Just Cry" by Mandisa
- "Just The Way You Are" by Bruno Mars
- "I Hope You Dance" by Lee Ann Womack
- "Many Rivers To Cross" by Jimmy Cliff
- "I Can See Clearly Now" by Jimmy Cliff
- "What A Wonderful World" by Louis Armstrong
- "Mama Said Knock You Out" by LL Cool J – my fight song!

YMCA LIVESTRONG PROGRAM

A year and some change into remission, I discovered that the YMCA had a Livestrong program – who knew? I didn't. It would have been great to take part in a wellness program six months into recovery, because exercise is key.

The Good Cancer

One night after a plasma infusion, I was sitting in the hospital lobby waiting for my mother to pick me up, and I began talking with an older couple. The wife asked me what type of cancer I had; from what I could garner, her husband also had cancer. I told the woman that I had Non-Hodgkins Lymphoma. As it turns out, their granddaughter was in remission from NHL. The husband looked at me, smiled, and as his eyes lit up he said, "You got the *good* cancer".

No cancer is good. They're all crap. This is the part where I say, "But" and negate everything I just said. But, from personal experience, Non-Hodgkins Lymphoma was not very invasive. I spent one day every three weeks getting treatment intravenously, no surgery, no months in the hospital. If Non-Hodgkins Lymphoma is the good cancer then yes, I had the *good* cancer.

Now That I've Taken My Wig Off

My wig now sits on a foam head by my apartment door, not as a reminder of how far I've come, because I believe that the worse thing I can do for my future is to live in the past. My wig is by my door honestly because it looks good and I put my headbands on it. Even though going through cancer twice was the worse time of my life, I tried to stay in the moment, not to enjoy the experience, but to be aware of changes to my body so that I could help my health practitioners help me. Oddly enough, I never thought about death or dying while dealing with cancer, maybe because both my liver cancer and Non-Hodgkins Lymphoma had very high cure rates.

What worried me the most was how my health was affecting my parents and sisters. My dad called me from Jamaica one day when I was at an all-time low. I tell you it was everything to hear him say, "Hello, sweetheart." I had nothing left to give to this disease, so I told him that if this treatment didn't work, I didn't think I could go through it again, which was probably not the best thing to say to your father. My poor Pop implored me to stay strong and to focus on getting better, because if I died it would literally kill my mother. Him, too, but my mom and I are like two peas in a pod. Let me be clear, I had no intention of giving up, but cancer and chemo have a way of making you question your very existence. They force you to start all over again physically, mentally, emotionally, and even financially.

I wanted to survive as much for my family as myself. I stayed in every agonizing moment of being weak and feeling miserable because I just wanted my f-ing life back. Once I got it back, I was gonna continue to be the best person I could be: laugh like I just heard the best joke ever, love my loved ones fiercely, smile and be fearless, and, yes, dance like no one was watching.

Diagnosis Checklist:

☐   Contact your health insurance company to see if you can get reimbursed for buying a wig.

☐   Ask your doctor for a wig prescription.

☐   If you will not be living at home during treatment, contact utilities and cable company to reduce your services – they may be able to freeze your account, which will save you money.

☐   Speak with your employer and/or clients about modifying your schedule.

☐   Collect the business cards from all of your doctors and make a master copy of them on one sheet of paper. Carry the paper with you to your appointments for check-ins and more importantly for your family and caregivers.

☐   Download Rx app for your pharmacy onto your phone.

☐   Go wig shopping early it'll make the transition to baldness easier and you may not have enough energy later.

☐   Do as much on the chemotherapy no-no's list as possible before treatment starts, for example get a manicure and pedicure and eat sushi.

Overnight Bag:

☐ PDA

☐ PDA Charger

☐ Phone

☐ Phone Charger

☐ Lock for bag

☐ List of meds

☐ NO medication – I was asked whether I had any meds on me and they were taken.

☐ Underwear

☐ Lotion

- ☐ Toothbrush

- ☐ Book

- ☐ Headphones

- ☐ Patience

- ☐ And don't forget your wig!!

MEDICAL GLOSSARY

Contrast:

http://www.mayoclinic.org/tests-procedures/mri/basics/what-you-can-expect/prc-20012903

R-CHOP:

http://www.cancer.gov/about-cancer/treatment/drugs/r-chop

Sepsis:

http://www.mayoclinic.org/diseases-conditions/sepsis/home/ovc-20169784

Chemoembolization:

http://www.cancercenter.com/liver-cancer/chemoembolization/

White Blood Cells/Absolute Neutrophil Count:

http://www.cancer.org/treatment/understandingyourdiagnosis/examsandtestdescriptions/understanding-your-lab-test-results

Ascites

http://www.webmd.com/digestive-disorders/ascites-medref

List of Prescriptions:

Notes: